Fire

Fireground Skills ~ Fireground Language

Fireground Survival Series

Firefighter Survival

Written by
Jim McCormack
with
Bob Pressler

P.O. Box 1852 • Indianapolis, IN 46206
http://www.fdtraining.com

Disclaimer

The recommendations, advice, descriptions and methods in this book are presented solely for educational purposes. The authors and publisher assume no liability whatsoever for any loss or damage that results from the use of any of the material in this book. Use of the material in this book is solely at the risk of the user.

Published by
Fire Department Training Network, Inc.
P.O. Box 1852
Indianapolis, IN 46206

ISBN 0-9719788-0-8

ABOUT THE AUTHORS

Jim McCormack has been a firefighter for 14 years and is currently with the Indianapolis Fire Department. He is the founder and president of the Fire Department Training Network, Inc.

As a member of the Fire Department Instructors Conference (FDIC) Advisory Board, and instructor with the FDIC Hands-On-Training courses, he continues to pursue firefighter training and education as the key to firefighter survival.

Bob Pressler is a 29-year veteran of the fire service who has recently retired from the City of New York Fire Department at the rank of Lieutenant. Before retiring, Bob had worked in 4 out of 5 Boro's in some of New York's busiest companies including Rescue 3 and Tower Ladder 44 in the Bronx and TL 157 in Brooklyn.

He is a member of the Advisory Boards and the Coordinator of the Hands-On Training (HOT) programs for FDIC in Indianapolis and FDIC West in Sacramento.

ABOUT THE FIRE DEPARTMENT TRAINING NETWORK

The Fire Department Training Network, Inc. is a not-for-profit training organization dedicated to training firefighters. The Network produces results-oriented, reality-based, training material for firefighters and fire departments, including a monthly newsletter and fire department training package, *FIRESCUE Interactive* and the *FIRESCUE Interactive Department Trainer.*

For more information on the Network visit:
www.fdtraining.com

FireNotes

Fireground Skills ~ Fireground Language

IN THIS FIRENOTE:

- FIREGROUND SURVIVAL TRAINING – WHAT'S THE BIG DEAL?
- PREPARING FOR FIREGROUND SURVIVAL
- MANAGING YOUR MAYDAY
- SCBA EMERGENCIES
- DISORIENTATION EMERGENCIES
- EMERGENCY ESCAPE TECHNIQUES
- FIREFIGHTER SURVIVAL TRAINING SESSIONS

WWW.FDTRAINING.COM

DEDICATION

We would like to dedicate this FireNote to you, the reader, for continuing to pursue a safer fireground through training.

Pass on your knowledge, it just might make a difference in the life of a firefighter!

Table of Contents

FORWARD

The most important point regarding firefighter survival training is that ***prevention is the key!*** If you can prevent the situation from developing then it won't ever become an emergency. If, however, you can't prevent the situation—*or the events were out of your control to begin with*—previous hands-on training with different firefighter survival skills just may save your life.

Reading a book, by itself, can not make you proficient at the survival skills necessary to keep you alive on the fireground. The emergency techniques described and shown in this FireNote must be practiced in the training environment under the supervision of a qualified fire department trainer who is familiar with them. **This FireNote reviews a number of firefighter survival skills but it can not and does not replace the required training that must be performed in order to become proficient at these skills.**

1

Firefighter Survival Training *What's the Big Deal?*

Let's face it, the most dangerous part of our job is still responding to fires. Unfortunately, with all the mandated training we must complete (EMS, Hazmat, WMD...) fire training just doesn't seem to be a priority—at least not where the scheduling is being done.

Another reason is the belief that, as a firefighter, you are immune to a fireground tragedy. The underlying attitude that *'It will never happen to me'* is sure to increase the chances that you—or someone working with you—may get hurt or killed. **Gone is Gone!**

The whole idea behind developing and implementing a fireground survival program is to decrease the chances of firefighters getting into emergency situations on the fireground (PREVENTION) and increase the chances of successfully resolving those fireground emergencies that do occur. In other words, when you or a member operating on the fireground encounter an emergency, prior training produces instinctive responses that result in the problem being solved and everyone going home.

A FEW MORE REASONS...

- Complacency
- Peer Pressure
- Less Fireground Experience
- Murphy's Law

Complacency

Nobody is immune to this! No matter how much one tries there will always be a level of complacency that exists. How many times have you responded to an automatic fire alarm, at the same building, that has turned out to be a false call? Has there ever been a fire in that building? What state of mind will you be in when you arrive at the building for an alarm call and see fire and smoke showing? **Complacency Kills!** If it doesn't kill directly then it is responsible for starting the tragic chain of events that does.

Peer Pressure

Peer pressure has no affect on you, right? WRONG! This is something else that nobody is immune to. Like it or not, attitudes and surroundings impact everyone. How many times have you not put on your SCBA, taken a hand tool, or even used your helmet chin strap, because you are the *oddball* in the group for doing it? Be honest. That's peer pressure at its best—the attitudes and actions of others somehow force you into doing things that you know shouldn't be done. Peer pressure, like complacency, kills.

Less Fireground Experience

Fires are down everywhere. Fewer fires means less experience which results in an increased chance of problems occurring on the fireground. Couple this with the continued loss of experience due to retirements and the end result is a younger, inexperienced, fire service. This inexperience is also an issue with new company and chief officers. Commanding fires takes as much experience as fighting them and requires a working knowledge of effective strategy and tactics. This working knowledge can only come from experience.

One proven way to deal with inexperience is to provide and perform continuous realistic training that develops basic fireground skills. This type of training results in an increased knowledge of the strategy and tactics that work—and those that don't. There is simply no other way to get it done. Mastering basic firefighting skills is a must. Training must be ongoing, challenging, realistic, and repetitive to maintain proficiency. This holds true for firefighters, engineers, officers, and chiefs. What you do or may have to do on the fireground should be second-nature. The fireground is not the place to find out that you've been a bit lax. There isn't time and firefighters lives will be affected by the actions and decisions made.

Murphy's Law

Remember, stuff happens! Even when you've done your best to prepare, things can still happen (Murphy's Law). A solid fireground survival program (firefighter survival, firefighter rescue, rapid intervention), coupled with a thorough knowledge of the basics of firefighting, just might get you through when all else fails.

Firefighter Survival Skills

Firefighter survival skills are self-rescue skills and techniques that firefighters may use to solve their own fireground emergency situations. These skills may be used in assisting others but they are primarily considered skills that will solve or prevent an individual fireground emergency.

The remaining portion of this FireNote will deal with the following firefighter survival skills:

- Firefighter Survival - The Foundation
- Managing YOUR Mayday
- SCBA Emergencies
- Disorientation Emergencies
- Emergency Escape Techniques
- Training

"In the heat of battle you don't remember very much. You don't think very fast. You act by instinct, which is really training. So you've got to be trained for battle so that you will react exactly the way you did in training."

Admiral Arleigh Burke, U.S. Navy

2

Preparing for Survival

A SURVIVAL ATTITUDE

As with most training issues dealing with firefighters, having the right attitude is the first skill. Guess what? It may happen to you, it could happen to you, and sometime during your career it probably will happen to you! A ceiling may collapse, you may become disoriented, your SCBA may malfunction, your water supply may be cut off, your exit may be cut off, or some other unexpected event may occur on the fireground that places you in a life or death situation with little or no time to react. Accept the fact that things can go terribly wrong on the fireground and take every opportunity to prepare for, and train for, the emergency before it happens.

KNOW YOUR EQUIPMENT

Personal equipment

Personal protective equipment (PPE) can and does prevent injuries—when it's used and used correctly. Every firefighter has a basic complement of PPE (bunkers, gloves, hood, helmet, flashlight, etc.) and it's their responsibility to make sure the equipment is maintained and working correctly. Sure, somebody else may actually perform repairs or maintenance, but it's the firefighter's responsibility to initiate the action.

Do you clean your SCBA face piece at the start of shift or after each fire? Do you wash your bunker gear or have it professionally laundered? Do you check your gear at the start of each shift?

Apparatus

Every seat on the apparatus is potentially yours. Every position on the apparatus may have to be performed by you. Don't wait until things go wrong to figure out what you would do if placed in the 'hot' seat during an emergency.

Know the apparatus. If it's an engine, how much water does it carry? How much does it pump? How much supply line does it carry? How much attack line? What size supply line? What size attack line? What type of nozzle is used on the primary attack line? How about the other attack lines? What additional equipment is carried on the engine and where is it located?

If it's an aerial ladder or platform, how tall is it? What's the effective reach? What size area is required to set it up? How many ground ladders are carried, where are they and what sizes are they? Where are the tools and equipment carried?

When you put your gear on the apparatus you're making a commitment to the other members, the public, and yourself, that you can perform the job.

Radio

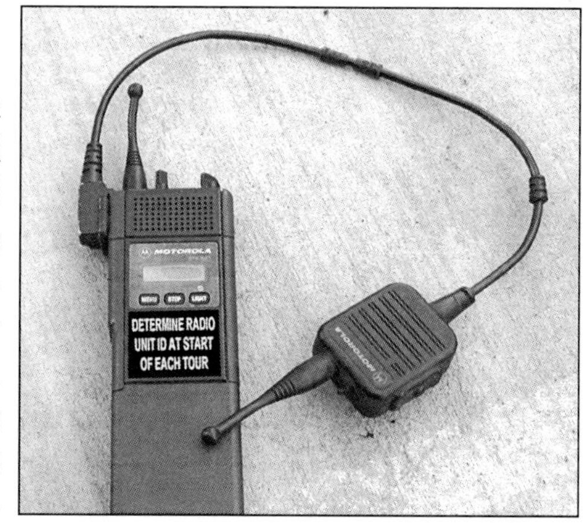

Every firefighter on the fireground should have a radio, period! Unfortunately, that's not reality. Every firefighter should know how to operate a radio, both during routine operations and fireground emergencies. How do you turn it on, change channels, change the battery, transmit messages and, if applicable, place it in alarm mode? Simple questions - yes! Routine, not during an emergency!

KNOW YOUR CREW

Who are you working with? Who are they working with? What are their skill levels? What are their attitudes? Are they having a good day or a bad day? Firefighting is a team effort and when the team isn't functioning then problems are more likely to occur. A working knowledge of the daily crew is essential to fireground success. Remember, you're a part of the crew and the above questions apply to you as well.

KNOW YOUR DISTRICT

A firefighter must be familiar with the response area. A working knowledge of the buildings, streets, water supplies and any problems in the area will add to the efficiency of overall operations. View the district on your way to work, while responding to and returning from runs, during preplanning and training sessions, and any other time you are exposed to it. Make a special note of changes or problems that will affect fireground operations. This knowledge of the district will minimize some of the variables when the response begins. There will be plenty of other things to think about.

PERFORM AN INDIVIDUAL FIREGROUND SIZE-UP

Every firefighter should conduct an individual size-up of the fireground. Whether you're the first-arriving officer who is establishing command or a firefighter on a later-arriving piece of apparatus, an individual fireground size-up is the first step in firefighter survival on the fireground.

Consider some of the following points while performing your individual size-up:

- Building footprint (length and width)
- Possible floor plan/layout
- Entry points
- Exit points
- Number of stories
- Presence of sub-floors or a basement
- Location of fire
- Extent of fire
- Progress of operating crews
- Location and nature of your assignment
- Any unusual conditions or hazards
- Any other pertinent information

Why? Always know what you're getting into and make sure you've considered what you might do if things go wrong. If the fire cuts off your exit what will you do? If you fall through the floor how will you figure out where you are? If you're knocked off of the wall or a hose line, how will you re-orient yourself?

Sizing-up the structure before you enter will help provide some of the information needed when things go wrong inside.

LEARN THE SKILLS FOR FIREFIGHTER SURVIVAL

There are many individual skills that may be required during a fireground emergency. The bottom line is that you must do whatever it takes to resolve the problem. Firefighter survival training provides a solid foundation so that, if faced with an emergency on the fireground, prior training causes you to *react* while you're thinking about how to get out.

3

Managing YOUR Mayday

MAYDAY! Sooner or later it just might happen to you. The big question is, ***are you prepared to deal with it?*** Consider the following scenario: upon arrival you encounter heavy smoke in a two-story double residence. The company you're with is assigned to search and rescue, the first due engine is in the process of making the hydrant and advancing the attack line—the search begins. The house is very cluttered and it's difficult to make progress. As you begin to make your way to the second floor the stairs give way and you're in the basement, your partner already made the floor and doesn't realize you're not right behind. The engine crew runs into a slight delay and the fire begins to gain the stairway to the second floor.

This is just one scenario. There are any number of situations that can lead to a firefighter MAYDAY. Consider the above scenario or think about a situation in your community that could produce a similar result. **What have you done to prepare for your own personal MAYDAY?**

Situations which would call for a MAYDAY:

- Imminent collapse
- Collapse has occurred
- Missing firefighter
- Unconscious/seriously injured firefighter
- Lost or trapped firefighter

Temporary confusion is not an emergency but it can lead to one. In the event that you're able to quickly solve the problem the following considerations may not come into play. Don't jump the gun but don't let tunnel-vision during a true emergency be the cause of a fireground tragedy. *Consider the following actions as some of the problem-solving steps to use during your MAYDAY:*

ORIENT YOURSELF

While this may be a difficult thing to do, take a few seconds to calm down and get your bearings. What's the status of your air supply? What were you doing? Were you near a wall? Advancing or following a hose line? Were you on the first floor, second floor, in the basement? Who was with you and where are they now? By performing a quick assessment of the situation you may be able to quickly solve it. Don't panic and begin moving aimlessly throughout the structure—your entry should have been systematic and your actions during this problem solving should be the same.

COMMUNICATE WITH YOUR CREW

There should be at least one other firefighter with you! Are you separated? Can you communicate? Constant communication during fireground operations is essential, especially among individual crews. If for some reason you are unable to contact your crew then it's likely they don't know your situation.

Fireground noise makes it difficult to hear, period! Radio technology doesn't help the situation. Does every member on the fireground carry a portable radio? If not, is there at least one radio per inside crew? Where do you carry your radio? In your chest pocket? Pants pocket? Do you use a lapel microphone? Can you hear everything that is being relayed over the radio? (Probably not!) Consider these things when trying to contact your crew inside a structure.

Consider the normal course of action in notifying your crew. The first thing that's usually done is simply calling out to the other crew members, through an SCBA. If the crew isn't real close then they probably can't hear you. A radio transmission may come next—if everyone carries a radio. What happens next if those attempts fail? One other option is sounding a tool. Make sure you activate your PASS device (see below) and then make as much noise as possible with the tool. ***Don't give up in trying to communicate inside the structure but don't let tunnel-vision prevent you from dealing with the real problem!***

ALERT COMMAND

This is the time that the radio you've always carried pays for itself—*if you use it!* All too often a firefighter in trouble waits until late in the stages of the situation before calling for help, Why? The easiest thing to do if the problem is overcome is to make another radio transmission declaring the problem solved. Whether you're injured and can't move or simply disoriented and beginning to troubleshoot the problem, notifying Command is an early step in the process. In the event that you're unable to solve the problem help is already on the way.

LUNAR

L Last known location

U Unit number/ID

N Name

A Assignment

R Radio equipped

Some of the more important things to relay to Command are: your unit number, the problem, your location (or last known location), what assignment you were performing and the status of your crew. It may be that your crew is unaware of the problem or they may be actively working to correct it. In any case it's important for accountability reasons to relay the information.

SOLVE THE PROBLEM

The above actions during your MAYDAY should take only a few seconds. Dealing with the problem should be your highest priority and the above actions are part of dealing with it. There are a number of skills that can be attempted to solve your MAYDAY and it's time to put them to the test. Use what you've learned, *improvise if you have to*, to find a way to solve your problem and get out. Help should already be on the way. *Don't omit the next step if your initial attempts to solve the problem don't succeed!*

ACTIVATE YOUR PASS

This means manual alarm mode–*it should already be turned on!* One of the first things that all firefighters need to focus on is the sound of an activated PASS device during fireground operations. The first thing that should be done when you hear a PASS device going off is to check and make sure it isn't your own. If it is then reset it! *How many times have you ignored a nearby PASS device? How often was it yours? How many times have you let an activated PASS device move right by you?*

During your personal MAYDAY, activate your PASS device to alert others that you need help. If the members of your department are trained and conditioned to proactively react to an activated PASS device then they'll know somebody is in trouble. You may only be a few feet from other firefighters but they won't know unless they hear you. Don't focus solely on solving the problem and become so exhausted or overcome that your PASS device must remain motionless for 25 to 30 seconds before it activates. That 30 seconds may allow the RIT to pass right by—they should be moving at a pretty fast clip because they're searching for a firefighter in trouble. *Why take a chance!*

SOLVE THE PROBLEM

This was discussed just before *Activating Your PASS* but is repeated here because it is critical to continue trying to solve the problem. Seconds count! Remain calm, orient yourself, know that help is on the way (you've broadcast a MAYDAY and activated your PASS), and systematically attempt to solve your problem.

IF YOU CAN'T SOLVE THE PROBLEM

If you're unable to solve the problem then do everything you can to make sure the RIT can locate you. Communicate with Command again. Communicate with the RIT if possible. Make sure your PASS is activated. Sound with a tool. At this point the success of the operation hinges on the RIT's ability to find you. ***Make sure you're prepared in the unlikely event that you find yourself in this situation!***

4

SCBA Emergencies

The only thing certain about SCBA emergencies is that you never know when they might happen. Unfortunately, sometimes the only experience one has with an SCBA emergency is the last one.

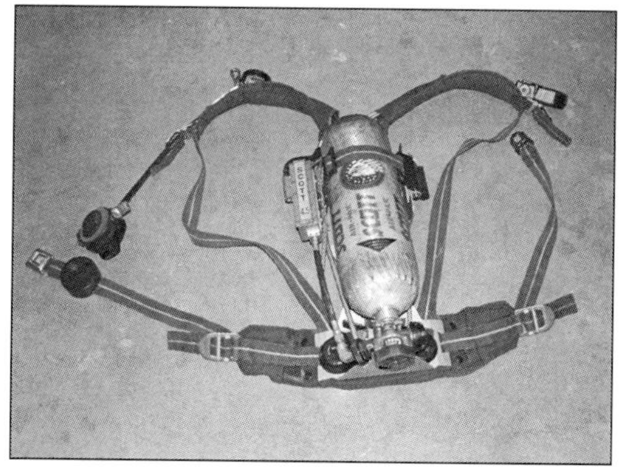

The self-contained breathing apparatus is still one of the most important pieces of protective equipment that a firefighter uses. Unfortunately, the amount of training time spent on SCBA is minimal. Every firefighter should take the time to become completely familiar with the SCBA used by the department, during routine maintenance checks as well as during fireground conditions. A thorough knowledge and understanding of the SCBA is essential for firefighter survival. It doesn't matter what brand or who the manufacturer of the SCBA is, it's every firefighter's responsibility to make sure they know everything there is to know about the SCBA they use.

DAILY MAINTENANCE AND INSPECTION

Seems simple, right? Every shift should begin with a basic review of the SCBA, how it's put together and how it operates. This check should also include proper functioning of the PASS device. Do you answer these questions when you check your SCBA? Is the high pressure hose firmly attached to the cylinder? Is the cylinder full (not almost full or that's enough, but full)? Does the cylinder gauge and the remote air pressure gauge match? Does the bypass valve work? Does the low pressure warning device work? Are the straps in good working order? Are they twisted? Do they adjust easily and properly? Are all

the hoses in good condition? Is the face piece clean and functioning properly? What else do you check?

DEVELOP AN SCBA DONNING ROUTINE

The importance of an SCBA donning routine is often overlooked, or not even considered, until there is an emergency. Whether the SCBA is donned from a jump seat, compartment, or case, it should be done in a consistent manner. A donning routine is a simple step-by-step method of putting on the SCBA. This routine is developed by consistently repeating the same actions in the same order. Over time these repeated actions become second nature and are performed without thought (instincts/reactions). During an emergency these *reactions* will allow you to continue dealing with the problem while focusing on your current situation.

Here's a Simple Donning Routine:

1. Don the SCBA harness using whatever method you prefer (jump seat, compartment, case).
2. Grasp both upper shoulder straps as high as possible on the shoulder and slide your hands down the straps to the buckles, making sure there are no twists in the straps. Continue down the straps, beyond the buckles, to the point where they connect to the backpack/harness. **Remove all twists.**

3. Slide both hands back up the straps to the buckles, grasp the loose ends of the shoulder straps and tighten them.

4. Grasp the waist straps as far back towards the backpack/harness as possible and slide both hands forward to the waist buckles, **removing any twists** in the straps.

Remove any twists in the straps during the donning routine and align the buckle in the center of your body. In an emergency you'll be able to find the buckle quickly.

5. Connect the waist buckle and tighten the waist strap making sure to keep the buckle in front, center, of the body. Make sure not to trap the shoulder straps under the waist strap. During an emergency the trapped shoulder straps may hinder loosening or removing the backpack.

6. If donning the face piece, properly don and secure the face piece making sure to get a proper seal with no twisted straps. Secure the helmet and hood.

   6a. If the face piece will be donned later, secure the face piece and regulator to prevent damage and debris accumulation.

7. When going on-air, turn on the SCBA cylinder (all the way) and confirm the pressure gauge reading with the cylinder reading (checked during daily/scheduled SCBA service check), connect the regulator and test the operation of the SCBA.

8. Activate the PASS device if not an integrated unit. (This step can be performed anywhere during the routine.)

NOTE: The entire donning routine should be practiced both with and without gloves. During an emergency firefighters will have to solve SCBA problems while wearing gloves. Proficiency must be developed during training.

SCBA SURVIVAL SKILLS

Confidence in the use of SCBA is an essential part of successful fireground operations. During initial firefighter training SCBA mazes and confidence courses are routine. Once basic training is finished, however, they are usually non-existent. The only way to remain proficient at dealing with SCBA emergencies, *even if you're using the SCBA every day on the street*, is to continuously challenge firefighters during training.

Consider the following...

Take a few minutes to review your knowledge of SCBA. Remember, if this were an emergency you wouldn't have time to stop and think about things!

- You're hung up on something and must remove your SCBA to free yourself...where are the strap adjustments located? How do they operate? Can you do it with a gloved hand?
- You're not getting enough air flow, seems like you may be running out, where is your bypass? What way does it turn to provide more air? What if that doesn't work?
- You suddenly hear a high-pressure air leak, which side is your tank cylinder valve on? Which way does it turn to close? Open? Have you ever had to exit a building while controlling the tank cylinder valve to conserve your air supply?

SCBA survival skills are those skills that a firefighter performs leading up to, or in, an emergency that help solve the problem and eliminate the situation. The solution may be as simple as a strap adjustment or may involve an emergency escape from the building.

Fireground emergencies involving the SCBA generally fall into one of four categories:

- Reduced Profile Maneuvers
- Entanglements
- Equipment Failures
- Out-of-Air Situations

Reduced Profile Maneuvers

All firefighters should be proficient at performing a reduced profile maneuver while wearing an SCBA. While this type of maneuver may be required to gain access to an area during normal operations, it will more likely be required when forced to retreat from an interior position due to rapidly-deteriorating fireground conditions when no other options are available. Three ways to reduce profile with an SCBA and get through a restricted opening are:

- Loosen/Adjust Straps and SCBA Harness
- Partial Removal of the SCBA Harness
- Complete Removal of the SCBA Harness

The method used depends on the size of the opening, the size of the firefighter, and the experience level of the firefighter. An essential and final step with any reduced profile maneuver is the rapid and correct re-donning of the SCBA Harness. This returns the firefighter's equipment to a pre-emergency condition preparing the firefighter to react to the next situation.

Loosen/Adjust Straps and SCBA Harness

Depending on the size of the opening, and any obstacles encountered, you may be able to simply loosen the straps (waist and shoulder) of the SCBA harness and get through. Loosening the straps allows the SCBA to slide and move on your back while working through the opening. A potential problem that may be encountered is that the SCBA may move enough to get 'hung-up' in the opening. Many times this can be corrected and overcome by simply backing up and shifting the SCBA by moving your body (rotating hips or twisting) while inside or approaching the opening and then continuing.

Knowing your flexibility and capabilities while wearing an SCBA, along with having an accurate ability to size up the opening/restriction, will play an important part in the success of this method.

Partial Removal of the SCBA Harness

Partial removal of the SCBA harness to reduce profile requires the firefighter to perform a few additional steps. The first step is to protect the regulator hose and the integrity of the regulator and face piece. As simple as this seems, not consciously performing this step could

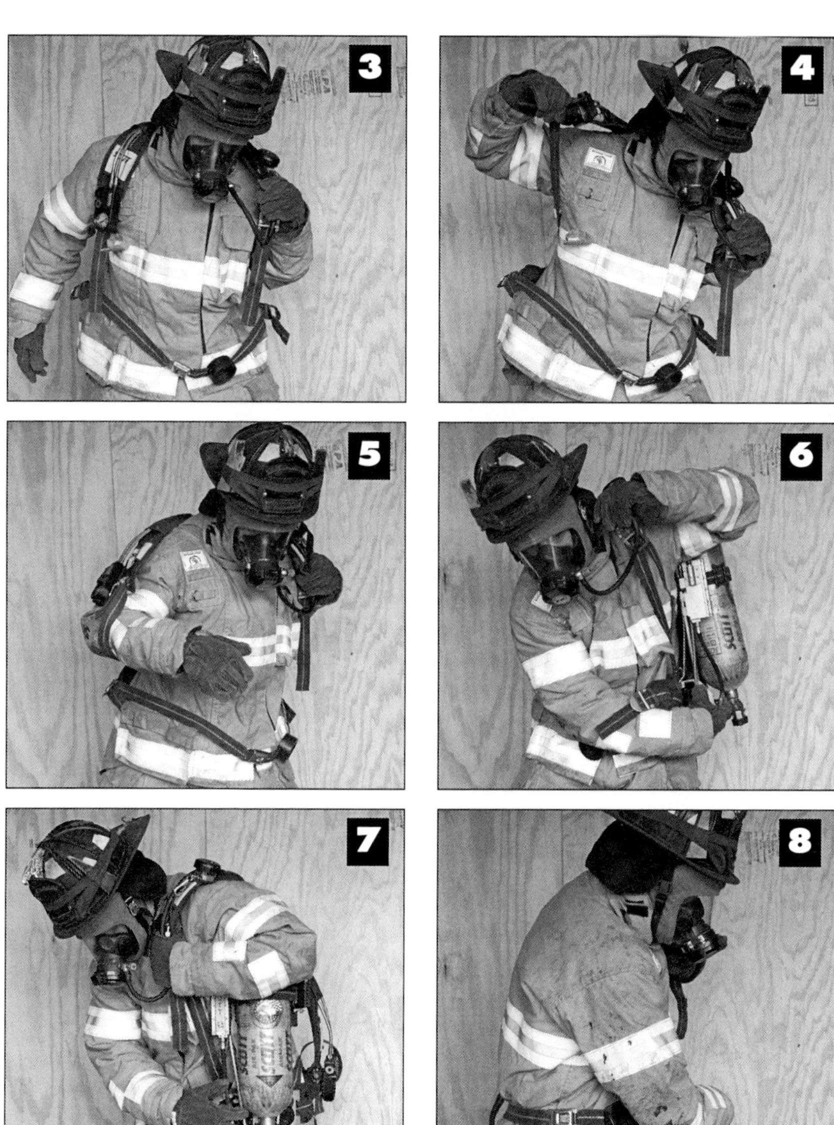

have tragic consequences. To protect the regulator hose, for mask-mounted regulator assemblies, firmly grasp and hold the regulator hose and SCBA shoulder strap on the appropriate side of the SCBA harness. (For SCBA with low pressure, belt-mounted, face piece hoses, protect the regulator and face piece integrity and make

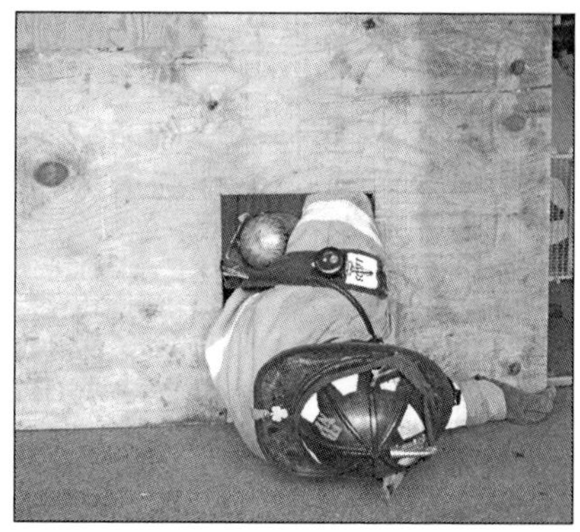

sure not to crimp the low pressure hose when moving through the opening.) Once the regulator hose is secure, loosen the waist strap to allow the unit to move freely on the turnout coat. *Only disconnect the waist strap if unable to loosen enough to allow movement.* Next, loosen the shoulder strap on the side opposite the regulator hose and remove it from the arm. Finally, while still grasping the regulator hose and shoulder strap, reach around with your free arm, grasp the SCBA cylinder where it threads into the harness and rotate the entire harness around to the side of your body until it rests underneath the armpit on the regulator hose side. This effectively reduces the profile of the firefighter (to a profile without an SCBA).

With the SCBA in the reduced profile position, cautiously re-approach and re-check the opening making sure to size up the conditions on the other side. When ready to proceed, work your way into and through the opening. **If you get stuck, don't fight it!** Stop and back up before making the problem worse. Adjust your positioning, if it will allow you to get through, and continue. If it isn't going to work then back out and use the Complete Removal Technique.

NOTE: Less is better when it comes to removing the SCBA harness to pass through a reduced profile opening. What's extremely important is to not expend valuable time and energy with repeated attempts that are not working. If the partial removal isn't successful the first or second time then change tactics—use the complete removal maneuver and then move on.

Key Points

- Always protect the regulator hose and face piece assembly.
- Only remove the waist strap if necessary. It's one less step when re-donning the pack on the other side.
- Remember, the top of the cylinder (head-first attempt) is the most common 'hang-up' point.
- Always size-up the opening before proceeding.
  - What are you dealing with?
  - Are there any obstructions in the opening?
  - Can you make the opening larger with a tool?
  - What's on the other side? Make sure to sound the floor!
- Re-don the SCBA after passing through the opening (includes connecting and adjusting all straps).

Complete Removal of the SCBA Harness

When the partial removal technique won't work the entire SCBA harness will have to be removed from the firefighter. This technique begins the same way as the partial removal and continues until the entire SCBA harness is removed. As always, the first step is to protect the regulator hose and face piece connection. Next, loosen the waist strap and disconnect it. Once discon-

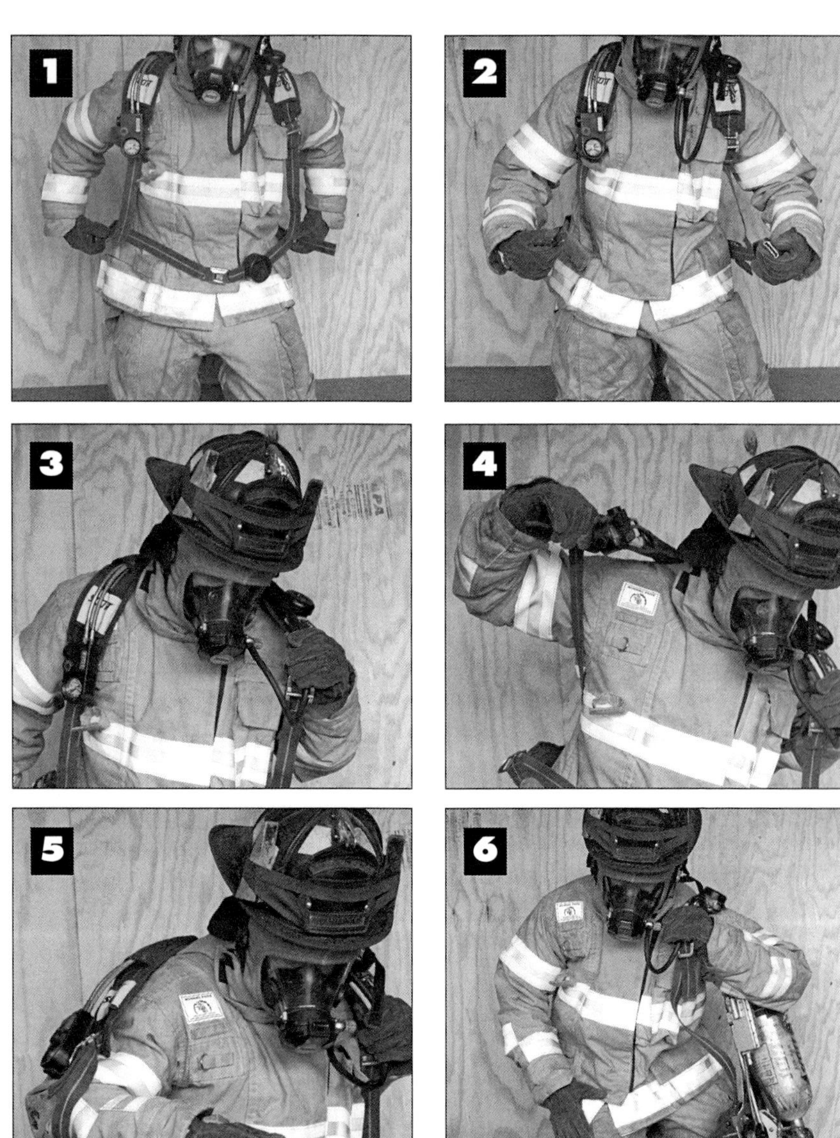

nected, loosen both shoulder straps and remove the strap opposite the regulator hose. While firmly grasping the regulator hose and shoulder strap, slip the SCBA off of the regulator-side arm and position it in front of you. Positioning the SCBA with the cylinder valve facing away from your body should maximize the working

length of the regulator hose (this applies to all SCBA units with the regulator hose coming over the top of the shoulder).

Once the SCBA is removed and your profile is reduced, re-approach the opening and recheck the integrity of the environment on the other side. While maintaining a firm grasp of the SCBA and, in particular, the regulator hose (failure may cause the unit to fall and be pulled away from you—jeopardizing the face piece seal) proceed through the opening, SCBA first. Once through, properly re-don the SCBA and continue.

Key Points

- Protect the regulator hose and face piece connection.
- Loosen and disconnect the waist strap.
- Loosen both shoulder straps.
- Remove the regulator-side shoulder strap last.
- Maintain a firm grasp of both the SCBA and the regulator hose throughout the maneuver.
- Re-don the SCBA after passing through the opening (includes connecting and adjusting all straps).

Head-First or Feet-First Through the Opening?

One question that often comes up is *"do you go through the opening head-first or feet-first?"* If you go through head first then you'll be able to pull yourself through using your hands. If you get stuck halfway when going through feet first then you'll be out of luck! The most important point to make is ***remember the objective*—get through the opening!**

ENTANGLEMENT EMERGENCIES

Dealing with an entanglement is a real possibility during any interior firefighting operation. Training to deal with entanglements goes a long way toward preparing firefighters to deal with these types of fireground emergencies. Entanglement emergencies may last only a moment or they may become a full-blown MAYDAY event for the firefighter, Command, the Rapid Intervention Team, and all other firefighters on the fireground. Being prepared to deal with the situation is the first step in solving the prob-

lem. What tools, if any, do you carry to deal with an entanglement emergency?

Consider all the possible entanglement hazards that may be encountered during every-day responses. An entanglement could be as simple as a curtain or window shade falling on top of an advancing firefighter or as complex as an entire suspended ceiling assembly coming down, either situation must be dealt with in a systematic and controlled manner in order to overcome the problem. Wires have always been a problem. A more recent hazard that's increasing daily, due to the popularity of the internet, is the number of computer cables and telephone wires being installed in both commercial and residential occupancies. Many times these cables are being installed in adhesive-backed strips applied overhead. These strips fail quickly under fire conditions.

Heating, Ventilation, and Air Conditioning (HVAC) equipment is also increasing the potential entanglement hazards that may be encountered. Take a few minutes to stop by a local HVAC contracting supply house, or a local home repair store, and look at all the potential sources of entanglement found with the various HVAC ductwork supplies available. What looks so forgiving, a five-inch aluminum covered oval duct, may become a potential firefighter killer during an actual fire. When the aluminum coating burns away the single piece of duct actually becomes an extremely long piece of coiled wire (basically a 'slinky') that could end your career.

Suspended ceilings are also becoming more popular as do-it-yourself homeowners try their hand at contracting. What was once considered only a commercial occupancy hazard, suspended ceilings must now be considered a likely hazard in residential homes as well. Two common locations for suspended ceilings in residential homes are finished basements and attics, although they may be

found anywhere. Suspended ceilings are made up of light-weight grid-work that is suspended on light-gauge wire. When exposed to fire these assemblies give way and come crashing down. Add the myriad of wire, conduit, insulation and ductwork that sits on top of these ceilings and the entanglement potential is enormous.

Firefighters who become entangled in wire or other debris during fireground operations face a true emergency situation. When dealing with an entanglement involving the SCBA you may by able to free yourself simply by repositioning (and subsequently removing the entanglement) and proceeding or you may have to perform a partial- or complete-removal of the SCBA harness in order to eliminate the hazard. Some fireground situations may also require you to use a pair of wire cutters to eliminate the hazard.

When faced with an entanglement on the fireground, STOP before making the problem worse. Four possible survival techniques that may help eliminate an entanglement emergency are:

- The 'Swim' technique
- Partial removal of the SCBA harness
- Complete removal of the SCBA harness
- Using wire cutters to cut away the entanglement

The 'Swim' Technique

When dealing with a simple entanglement, or the possibility of one, it may be possible to use the 'Swim' technique to eliminate or avoid the hazard. If an entanglement occurs, STOP! Many firefighters try to force their way through an entanglement and usually make a bad situation worse. Once you recognize the entanglement back up, while lowering your body position, to try and remove the problem. If you're able to remove the hazard the 'Swim' technique may allow you to continue through the area without further entanglement.

To perform this technique, lower your body position and extend an arm out in front of you. While slowly proceeding under the hazard, sweep (performing a simulated swimming stroke) your extended arm up and over your body while simultaneously rolling the

SCBA cylinder down toward the ground away from the hazard. Your arm may keep the potential entanglement hazard up and off of your body while allowing you to proceed through the area without further problems. If this technique doesn't work, or you must remove an entanglement before proceeding any further, one of the other methods will have to be used.

Partial Removal Technique

This technique is similar to the partial removal of the SCBA harness to reduce profile. Once the shoulder strap is removed, and while protecting the regulator hose and face piece, turn your body toward the remaining shoulder strap and face into the SCBA (when reducing profile

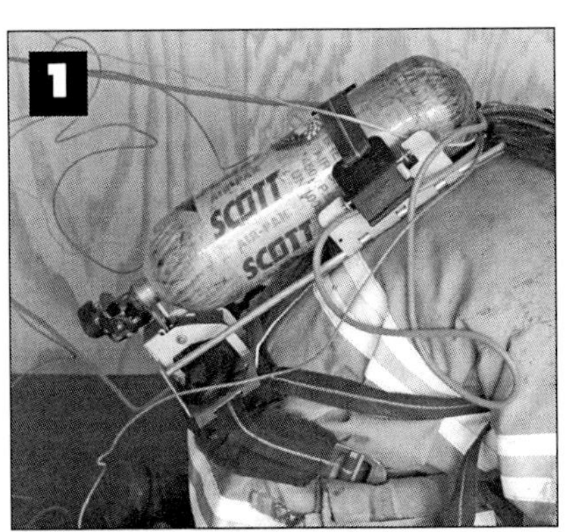

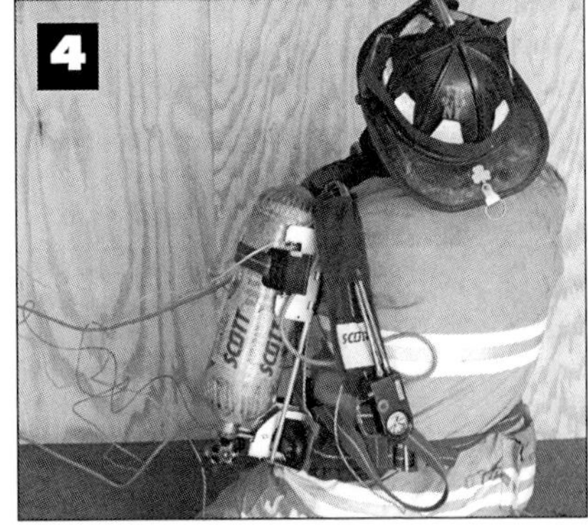

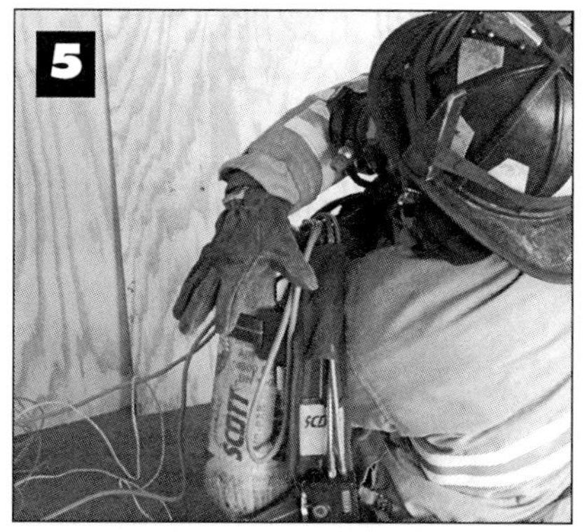

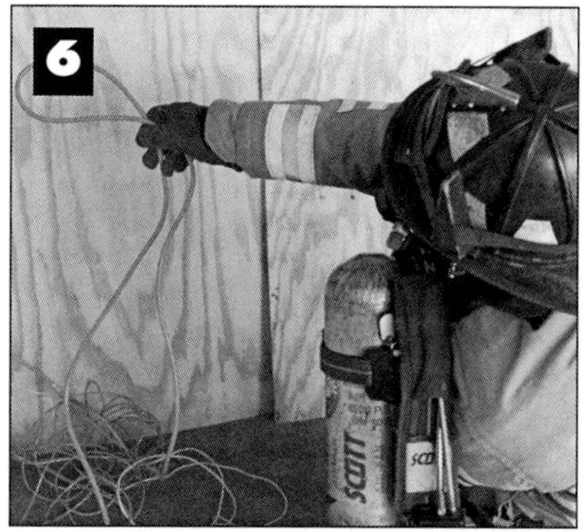

you normally bring the SCBA around your body). Attempting to pull the SCBA around toward the front of your body may complicate the entanglement. Once you've turned around to face the SCBA harness, work to remove any entanglement by sweeping your free arm around the cylinder and harness to find the hazard. Once the entanglement is removed, be sure to clear it from the immediate area so it doesn't become an additional problem. Next, move away from the hazard area and re-don the SCBA. In the event that a partial removal doesn't work then proceed with the complete removal technique.

Complete Removal Technique

The complete removal technique to remove an entanglement hazard is also similar to the one used to reduce profile. The difference, as with the partial removal technique, occurs when removing the shoulder straps. After removing the first shoulder strap begin to turn your body toward the remaining shoulder strap to face into the SCBA. While turning, slip your remaining arm out of the second shoulder strap. Remember, maintain a firm grasp of the SCBA harness and regulator hose to avoid problems with the face piece connection. Once you've removed the SCBA and are facing the harness,

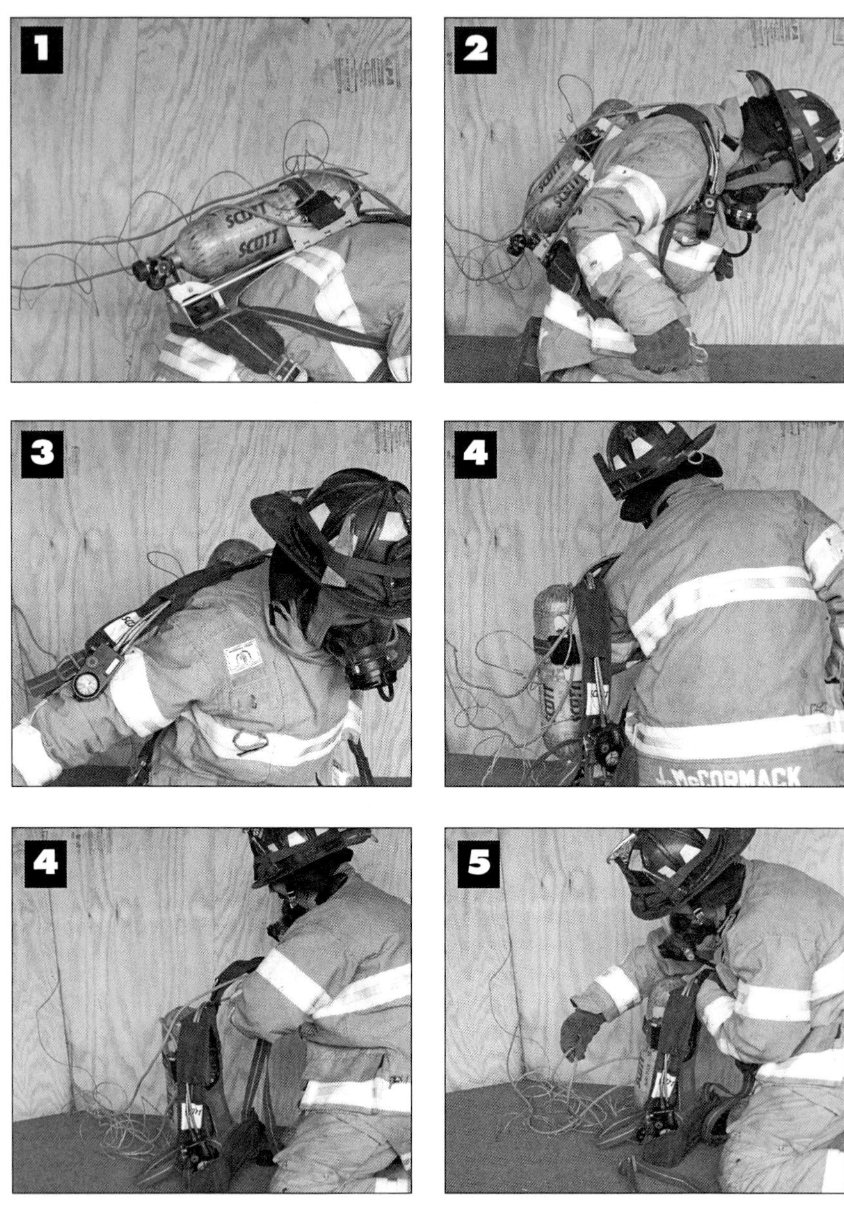

work to remove any entanglement by sweeping your arms around the cylinder and harness to find the hazard (making sure to keep a grasp on the SCBA during the process). Once the entanglement is removed, be sure to clear it from the immediate area so it doesn't become

an additional problem. Next, move away from the hazard area and re-don the SCBA.

Wire Cutters

There may be entanglements that can't be completely resolved by one of the above methods. In these instances you may be able to use a pair of wire cutters to eliminate the entanglement. Two obvious questions that must be asked are: Do you carry a pair of wire cutters and have you ever trained on removing and using them during limited visibility conditions with gloved hands (the conditions likely to be encountered during a fireground emergency)?

If you're unable to eliminate the entanglement hazard using the above techniques then continue communicating and make sure to activate both your PASS device and MAYDAY (if you haven't already). Assisting Command, the Rapid Intervention Team, and any other operating units in locating you will be instrumental in the success of your emergency.

Entanglement Training Sessions

A few simple training sessions are all that's needed to reinforce entanglement hazards and show the value of an inexpensive pair of wire cutters.

Create a simple, re-usable, training prop that includes wires, insulation, ductwork and suspended ceiling material. Create a way to suspend the prop above a room or training area that allows firefighters to move into, and under, the prop. When the firefighter(s) is under the prop release it onto him. The objective of the exercise is for the firefighter to become free of the entanglement and move out of the area. For advanced sessions include enough material so the firefighter must use a pair of wire cutters. BEWARE, the objective is to create a challenging but successful outcome, however, if a firefighter is unable to become free then include MAY-DAY and RIT training scenarios.

SCBA EQUIPMENT FAILURES

SCBA failures are rare but they do happen! When faced with an SCBA failure during fireground operations prior training will increase your chances of successfully dealing with the situation. SCBA equipment failures may fall into one of the following categories:

- Strap, buckle, harness failure
- Face piece failure
- Hose failure
- Other failures

While it's not possible, or intended, to make you an SCBA repair technician during this section it is possible to heighten your awareness of potential SCBA problems and some of the possible solutions. Some of the problems, and solutions, may seem extreme but remember never say never. Murphy lies in wait on every fireground!

Strap, Buckle, and Harness Failures

While the failure of a strap or buckle isn't a true emergency, by itself, the problem may 'snowball' into other more serious problems. Broken or loose straps (and especially a waist strap that's not connected) may lead to entanglement problems. A harness failure may result in the cylinder separating from the harness. If failure of any of these components occurs during fireground operations, take the time to solve the problem, which may include leaving the building.

Face Piece Failures

The SCBA face piece provides protection from the interior environment by allowing you to breathe and by providing a thermal barrier. Possible failures of the face piece may include a broken lens or a damaged or broken adjusting device (straps, netting). Because the face piece is often carried in an exposed position, hanging from the regulator hose while not in use, it is often vulnerable to damage.

If a face piece fails and compromises your air supply the only complete solution is to get to the outside and resolve the problem. Getting there is the issue. If an adjusting device fails you'll be able to hold the face piece against your face and exit. If the actual lens, lens assembly, or any other part of the face piece that protects you from the interior environment fails then you should get as low as possible to the floor and make your exit. By getting low you'll be exposed, theoretically, to the least toxic conditions and you may even find a layer of cleaner air. While staying low, you should filter the environment by covering the exposure point (crack, hole, etc.) with a glove, your arm, towel, piece of clothing found, or anything else that will help filter the smoke out while allowing you enough air to breathe. Conditions will be tough, to say the least, but if you don't attempt to get out then your situation will definitely not improve. Don't forget to hit a MAYDAY during this true emergency.

Hose Failures/High-Pressure Air Leaks

Any SCBA hose failure, or compromise, during interior operations represents a true emergency. While a low pressure hose failure will deplete the air supply slower than a high pressure hose failure, both will deplete the air supply and could leave you with a complete out-of-air emergency (see below). Again, as with an SCBA face piece failure, the only solution is to get out of the

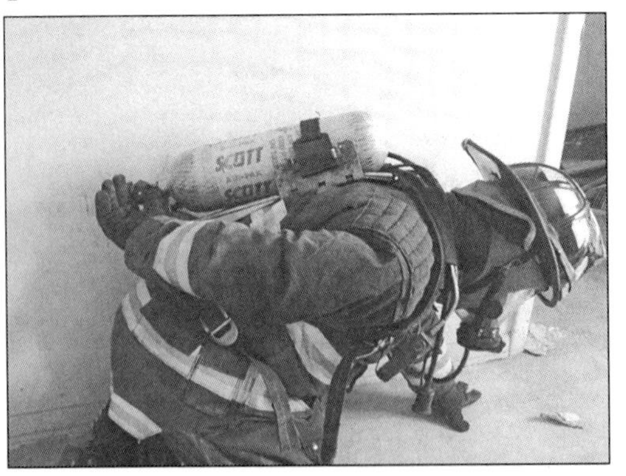

environment and solve the problem.

One of the most difficult things when dealing with an SCBA hose failure may be recognizing the failure. If a low-pressure hose fails it may not be noticeable because your air supply may not appear compromised and your breathing remains normal. Sound may be the only indication, and you may not even be able to hear a low pressure leak during interior operations. Constant awareness of your air supply, what you started with, what you're using, and what's left, may help you identify this type of problem.

A high-pressure air leak should be easily noticed (sound alone) and must be immediately controlled and corrected. A broken or compromised hose is the most likely cause of this type of situation, but don't rule out something else, including a blown O-ring (due to a loose connection between the cylinder and the high-pressure hose prior to opening the cylinder valve). The only true solution to control this type of leak and let you get out is to control the air supply by operating the cylinder valve.

- Which way does your SCBA cylinder valve turn to open/close?
- How many turns does it take to fully open/close the cylinder?

Control any high-pressure leak immediately and get out of the environment. Your breathing may or may not be affected by the high-pressure leak. If air is still making it to your face piece then you can control the air supply with the cylinder valve. If the air is not making it to your face piece, basically an out-of-air situation, then you'll have to resort to anything you can to get out (see cylinder breathing, face piece failures, out-of-air emergencies).

USING DIRECT CYLINDER AIR (CYLINDER BREATHING)

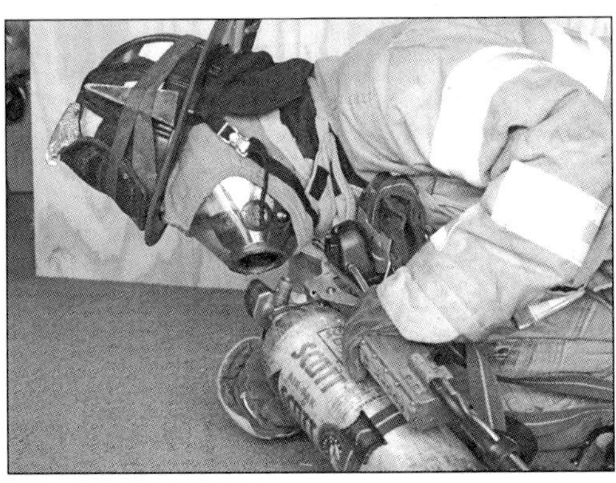

If there is still air available in the cylinder then you may be able to use it on your way out. Cylinder breathing, an extreme solution, will allow you to use some of the remaining air to get out of the area.

To use the remaining cylinder air you'll need to remove SCBA harness from your back and the high-pressure hose from the cylinder threads. **Remember**, this extreme solution uses direct, high-pressure, air so you'll have to use a 'blow-by' approach to get a breath of air. Position the cylinder opening near your mask opening (or near the opening of the low-pressure hose for belt mounted face piece connections) and take a breath by quickly turning on the cylinder enough to allow air into your face piece. Turn off the cylinder valve after each breath and filter out the environment while making your way out.

Other SCBA-Related Problems

There may be other SCBA failures during fireground operations. These failures, which may include internal operating components of the SCBA, should be handled using one of the above techniques.

One problem that may be encountered is a failure of the SCBA regulator. While this is an uncommon situation it is a possibility. One step that should be performed when your air supply appears to be compromised is to open the bypass valve. If opening the bypass valve appears to correct the problem then there could be an internal failure. Communicate your problem and exit the building to have the problem corrected.

Key Points

- Alert Command, other operating companies and your crew by hitting a MAYDAY before it's too late.
- SCBA emergencies can occur at any time.
- Extreme emergencies require extreme solutions, getting-out-alive is your number one priority.

OUT-OF-AIR EMERGENCIES

Have you ever thought about what you'd do if you ran out of air inside a smoke-filled building? Have you ever practiced dealing with this type of emergency? Training is the

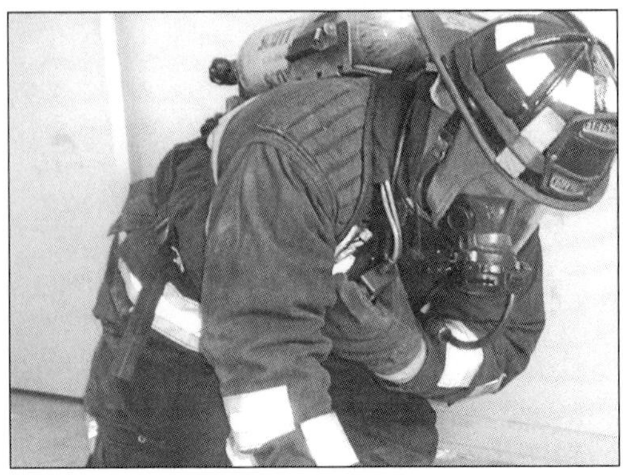

Step 1: Activate your PASS!

key to dealing with out-of-air emergencies on the fireground. Whether the emergency is a result of an equipment failure or an exhausted air supply, the result is the same, survival depends on getting out of the environment.

When faced with an out-of-air situation on the fireground there are a couple of solutions. The first is to immediately make a buddy breathing connection (if available), with a nearby firefighter, and then get out. The second solution, if no buddy breathing option is available, is to get out!

Assistance from a nearby firefighter (buddy-breathing)

Is your SCBA equipped with a buddy-breathing connection? Have you practiced making the connection?

This option, buddy breathing, will only work if the SCBA is equipped with a 'buddy-breathing' connection. Many newer SCBA are equipped with this option and it can be used to get an out-of-air firefighter to the outside. SCBA buddy breathing connections are currently not compatible between manufacturers so this option won't work unless both units are the same. A universal connection should be available shortly.

Proficiency in performing buddy breathing must be developed during training under realistic conditions. Connecting two SCBA units with a buddy breathing connection is difficult, even under favorable conditions, so prior training must be done. Once the connection is made it's critical for both firefighters to get out. The air supply that's available is being shared by both firefighters and it won't last long. Don't forget to communicate the problem and have assistance on the way in to help.

Getting to the outside

When faced with an out-of-air emergency the bottom line is that you must get out. Depending on the building, getting out may mean moving to a safe area (where SCBA isn't required) or it may mean getting completely out of the building. In either case, survival depends on getting enough clean air to breathe (survive) while you find an exit.

The cleanest air available, if any, will

be as low to the floor as possible. When you've run out of air and must perform an emergency escape, get as close to the floor as possible and filter each breath with your gloved hand. In order to breathe you'll have to remove a mask-mounted regulator from the face piece or a belt-mounted low-pressure hose from the regulator connection. With mask-mount openings, cover the opening with your glove (taking a slow-deep breath when needed) while keeping the opening as close to the floor as possible. With low pressure face piece hoses you can filter the air by covering the hose with a gloved hand or you may be able to tuck the hose inside your turnout gear. The bottom line is to filter any air coming into the mask, keep the mask as low to the ground as possible and keep searching for an exit.

Trapped: Out-of-Air

In the event that you're trapped and out of air the only possible solution is to filter the air as much as possible and wait for help. Activate your PASS device, if you haven't already, and try and slow your breathing (see below) to extend your air supply as long as you can. This may be your only option.

Air Conservation

An important air consumption drill that should be performed by every firefighter is to find out how long they can breathe off of an SCBA cylinder once the low air alarm has sounded. The drill can be done with a full cylinder (takes longer to get to the low air alarm) or the cylinder can be bled off until it's close to the pressure required to set off the low air alarm. To conduct this drill, put on all personal protective equipment and begin breathing off the SCBA. Bring your heart rate up to a working rate (use a stationary bike, stairmaster, treadmill, or simply exercise and move until your heart is working). Once you're at a working rate and the low air alarm starts to sound, stop and sit or lay down and start the clock. Try and relax and slow your breathing while you breathe off of the SCBA. If you were trapped inside a building and had to wait for help to arrive, the knowledge gained by performing this drill may assist you in relaxing and extending your remaining air supply. ***Extreme training for extreme situations!***

5

Disorientation Emergencies

Firefighter disorientation—it really does kill firefighters! Nobody ever expects to get lost inside a building but it happens. Conditions change, sometimes faster than you can imagine, and a task as simple as advancing down a hallway can become a fight for your life to find a way out. Too many firefighters have the attitude *it won't happen to me and if it does it won't be any big deal finding my way out!* **Wrong!**

Disorientation on the fireground is a scary thing, it's not something that happens often and it's really tough to simulate in training. Disorientation may result from any number of fireground events. A few of the common causes are:

- Loss of reference point (wall, hose, rope) for any reason
- Fall due to partial collapse of structure (into basement, through roof)
- Collapse of structural components (hit by ceiling, wall)
- Rapidly deteriorating interior conditions

A firefighter disoriented (lost) inside a building is a fireground emergency that, if unable to be resolved by quickly finding a reference point, will usually lead to an emergency escape or a tragedy.

DEALING WITH DISORIENTATION

When a firefighter becomes disoriented it's critical to STOP and gain control of the situation. There's no magic timeframe when it comes to how long it will take to gain control it's just essential that you do it. Failure to gain control will begin (or continue) a panic cycle that may be difficult, if not impossible, to stop.

Do a quick size-up of what just happened and what your current situation is. Where is your crew? Can you communicate with them? Was there a collapse? Did you get knocked off of your reference point? Did you simply lose your sense of direction inside the building? Try and relate what just happened to where you may be. Did the ceiling come down and knock you off a reference point? Did the floor collapse and drop you onto the floor below? Are you in a basement or crawlspace? Giving thought to what happened and where you might be will help you orient to the current surroundings. Once you've gained control of yourself it's time to find a reference point so you can figure out your next move.

When you're dealing with disorientation on the fireground consider the following options:

- Communicate with crew/Command
- Search for a reference point
  - Wall
  - Hose line
  - Building features
- Perform an emergency escape (see Emergency Escape Techniques, Chapter Six)

Communicate with Crew/Command

Communicating with your crew may be the easiest way to find out where you are. If you were operating together then simply call out for them. The longer you wait to do this the further they may get away from you, especially if they don't know that you're disoriented. Radio communication is another important point. If you're disoriented and by yourself, unable to find a reference point, don't forget to let somebody know about your situation (see Managing YOUR MAYDAY). Don't forget your PASS device!

Search for a Reference Point

Once you've communicated your problem it's time to deal with it by finding a reference point. A reference point is anything that will help you become oriented and continue with your mission or get out (depending on the situation). The two most common reference points, that will assist you in getting out of the building, are a wall or a hose line.

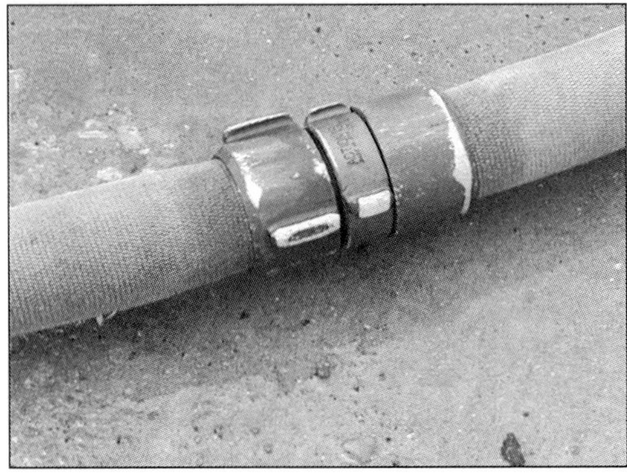

When moving to find a reference point it's important to keep some sense of where you are, and where you've been. Don't just aimlessly wander throughout the area. Begin from a certain point and move in one direction (as best you can) for a certain distance. If you don't find a reference point, consider moving back to the starting point and searching the opposite way (180°). If no reference is found then consider searching perpendicular to your original search direction, both ways if needed. The footprint of the building will be helpful in determining how far to search in each direction. In residential occupancies the distance between outside walls may only be 30 to 40 feet. In large commercial occupancies the travel distance to safety may be much greater.

In larger areas, after determining a direction to search, consider throwing an object out in front of you to see if there is a wall up ahead. If there isn't anything available, and you have a short piece of rope, consider swinging the rope out in front of you to contact a wall or object. Finding anything that may indicate where you are will be helpful.

Following a Wall

When you find a wall, follow the wall until you encounter a window or door. When searching for a door or window make sure to sweep the wall high enough to feel the windowsill. Often times a window is missed because the searching hand didn't sweep high enough up the wall. How far you have to travel along a wall until you encounter a door or window will depend on the type of building you're in. Performing a pre-entry size-up will give some indication of how far this may be. If you encounter a window, and conditions are still tenable inside, open the window and communicate your location. If conditions require you to get out immediately then perform an emergency escape (see Emergency Escapes Techniques).

If you encounter a door you must decide whether to pass through it or pass by it. Again, a building size-up prior to entry may provide valuable information at this point. Whatever your decision, you'll need to continue along a wall until you reach safety. If conditions require you to get out immediately then perform an emergency escape (see Emergency Escapes Techniques).

Following a Hose Line

Following a hose line provides a more direct link to safety than following a wall. The hose line was stretched from either an engine or a standpipe system. Following the hose to the engine will get you out of the building. Following the hose to a standpipe connection should lead you to a safe area. This is why following firefighting basics (hooking to the floor below) is so important.

The first step in following a hose line to safety is locating a coupling, or the nozzle, to determine which direction you're heading. If you follow the line and run into the nozzle first then the exit is the other way. If there are no firefighters operating at or near the nozzle that are able to help, then communicate your position, turn around and follow the hose line out. If you follow the line and run into a coupling first then you'll have to determine which direction you're heading by studying the couplings.

Male couplings are one piece and have long lugs on them. Female couplings have two pieces, one piece is long and smooth and the other piece is short with lugs. The lugs on the female coupling are much shorter than those found on the male coupling. The smooth piece of the female coupling is closest to the hose and the short piece with lugs threads onto the male coupling. Once you determine which coupling you encountered first, male or female, you'll be able to determine the way out.

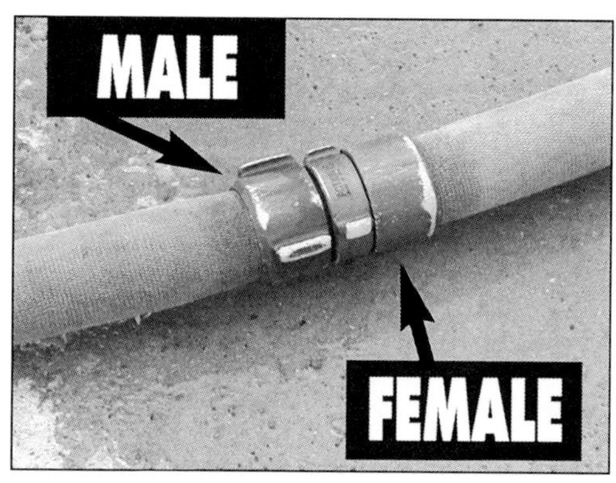

Think about how the hose line is stretched, male discharges and couplings point into the building. Female couplings point out of the building. If the first coupling encountered while following the hose line is a male coupling then you're headed toward the nozzle (deeper into the building). If, however, the first coupling encountered is a female coupling then you're headed toward safety (standpipe or outside).

If a reverse stretch is used on the fireground it should be communicated to all members. A reverse stretch (where a double male and double female adapter would be required) will reverse the coupling pattern described above. This reverse stretch may be possible with 2 ½ - or 3-inch line. While this type of reverse stretch is unlikely, anything is possible.

Always Keep Contact With the Hose

When following a hose line it's important to keep in contact with the line at all times. The hose line may go over or under obstacles, be wrapped around obstacles, be wrapped onto itself or in loops, or be

wrapped or twisted with other hose lines (spaghetti). Any obstacle encountered could cause you to lose direction if you let go of the line. Keep constant contact with the hose line. When encountering an obstacle, keep one hand on the hose line in front of the obstacle and put the other hand on the hose line beyond the obstacle. Slide both hands together so you're sure it's the same hose line. **Don't lose contact with the line.** If you must move an obstacle, place the hose line under a knee or between your legs but don't lose contact with it. Quickly recheck couplings when encountered to ensure that you're moving in the right direction.

Using Building Features

If you're disoriented in a large area, and/or you haven't been able to find a wall or a hose line, consider using some of the building components to provide direction. Many building features follow a certain pattern that may be helpful when nothing else is working.

Expansion joints

Expansion joints are common in large concrete slabs. Concrete slabs are common in many of today's larger

commercial structures. Expansion joints run in a straight line, both in-line and perpendicular, between columns and will usually lead to a wall. When trying to find a way out this building feature may lead the way.

Columns

As with expansion joints, columns usually follow a set pattern. In fact, many columns are set on a concrete footer and the footer is separated from the main concrete slab with an expansion joint. Columns, as with expansion joints, may provide assistance in getting out.

Column shape may also be helpful when no expansion joints are present. The face of a square column or I-beam (in any direction) will usually point toward another column or a wall. Round columns won't provide as easy a reference but may be accompanied by an expansion joint that will indicate direction to another column—or a wall.

Suspended ceiling grids

Another feature that may help you get out of an area is a suspended ceiling grid. If interior conditions allow

you to use this feature then consider how it is put up. These ceilings are set on grids that run between walls. If the ceiling is still intact you may be able to determine which direction leads to a wall. If the ceiling has come down you still may be able to determine a direction (many grids come down in large sections but remain somewhat intact) by following the grid.

FLOOR JOISTS

Floor joists (and many rafters) will also follow a certain pattern when used in construction. Floor joists will run between walls. If all else fails, it may be possible to open up the floor and determine the direction of the joists. Remember, opening a floor isn't an easy task, but in a last ditch effort it may help you find a way out. Some of today's light-weight components (truss floors, wooden I-beams) run long distances. Conventional frame construction (dimensional lumber) will have span limitations. In older ordinary construction, floor joists usually run parallel to the street (as a rule: follow the joists to a wall, cross the joists to the front or back of the building) ***Know your construction before you go in.***

6

Emergency Escape Techniques

Sooner or later it will happen! The department hits a working fire, you're performing interior operations and the situation gets out of control. The smoke drops to the floor, the heat becomes untenable, you and your crew are forced to the ground and it's just a matter of time before the room lights up. Where was the last door or window? You made entry up the interior stairs but you're too deep in the room and not likely to make it back to the door under these conditions. What are your options? When was the last time you reviewed and practiced emergency exits during interior fireground operations?

Firefighter safety and survival starts with you! By being aware of the potential for emergency egress, and being proficient at how to do it, your chances for success are drastically increased. Prevention is still the key but stuff happens so you'd better be ready to deal with it. Under emergency conditions your 'reaction' will be in direct proportion to the amount and type of training you've done.

Emergency escape should be a last resort. As stated earlier, preventing the situation is the first objective. When prevention doesn't work then consider leaving the area the way you came in. If your exit is cut off then consider finding a safe area by closing a door between you and the fire. Breaching a wall to a safe area may also buy you enough time to allow help to reach you. When all else fails and you're forced to leave through a window then perform a quick size-up of the situation and available options. Can you buy enough time by hanging on the outside of the window until the condition passes or do you have to get completely out? Consider what floor you're on and whether there are any ground ladders (or aerials) available to assist you. Emergency escapes from upper floors are extreme measures for extreme situations.

Here are some of the conditions that may result in having to perform an emergency escape during an interior fireground operation:

- Caught by rapidly deteriorating conditions
  - Flashover
  - Backdraft
- Sudden collapse
- Loss of water supply

- Inadequate flow
  - Improper pump pressure
  - Improper line selection
- Lack of ventilation
- Improper ventilation
- Disorientation

When faced with any of the above situations prior training and preparation in emergency escape techniques may help you get out. Emergency escapes may fall into one of the categories listed below.

- Follow a hose line to safety
- Rapid location of a door to a safe area
- Wall breach
- Emergency Window Exits
  - Straddle sill/hang
  - Hang and drop
  - Head-first ladder slide
  - Rope slide

> **Note:** When making an emergency escape by following a wall you'll use whatever option presents itself first, door or window. Each option is presented separately in the following discussion to highlight the differences.

A NOTE ABOUT EMERGENCY ESCAPE TRAINING:

All training should be done under the guidance of a qualified fire instructor. All safety measures must be in place when performing any training skills. All training that involves emergency exits from upper floors should include life safety lines and harnesses. **While the risk of injury still exists—the alternative to not performing this type of training is far more costly!**

FOLLOWING A HOSE LINE

If you encounter a hose line during an emergency escape from the building then follow the hose line to safety. (The following paragraphs on following a hose line are repeated from the Disorientation chapter.)

The first step in following a hose line to safety is locating a coupling, or the nozzle, to determine which direction you're heading. If you follow the line and run into the nozzle first then the exit is the other way. If there are no firefighters operating at or near the nozzle that are able to help, then communicate your position, turn around and follow the hose line out. If you follow the line and run into a coupling first then you'll have to determine which direction you're heading by studying the couplings.

Male couplings are one piece and have long lugs on them. Female couplings are two piece, one piece is long and smooth and the other piece is short with lugs. The lugs on the female coupling are much shorter than those found on the male coupling. The

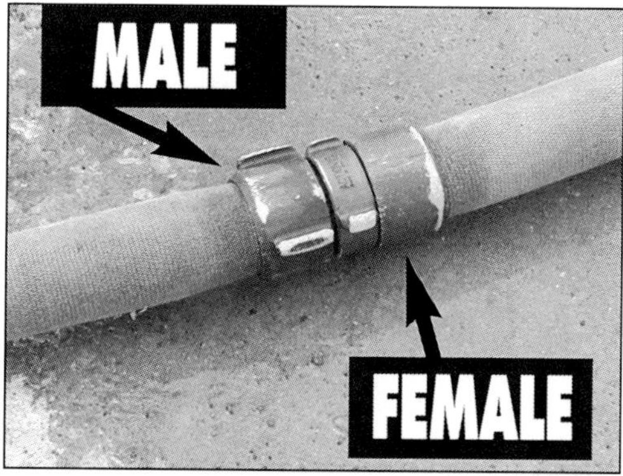

smooth piece of the female coupling is closest to the hose and the short piece with lugs threads onto the male coupling. Once you determine which coupling you encountered first, male or female, you'll be able to determine the way out.

Think about how the hose line is stretched, male discharges and couplings point into the building. Female couplings point out of the building. If the first coupling encountered while following the hose line is a male coupling then you're headed toward the nozzle (deeper into the building). If, however, the first coupling encountered is a female coupling then you're headed toward safety (standpipe or outside).

If a reverse stretch is used on the fireground it should be communicated to all members. A reverse stretch (where a double male and double female adapter would be required) will reverse the coupling pattern described above. This reverse stretch may be possible with 2 ½ - or 3-inch line. While this type of reverse stretch is unlikely, anything is possible.

Always Keep Contact With the Hose

When following a hose line it's important to keep in contact with the line at all times. The hose line may go over or under obstacles, be wrapped around obstacles, be wrapped onto itself or in loops, or be wrapped or twisted with other hose lines (spaghetti). Any obstacle encountered could cause you to

lose direction if you let go of the line. Keep constant contact with the hose line. When encountering an obstacle, keep one hand on the hose line in front of the obstacle and put the other hand on the hose line beyond the obstacle. Slide both hands together so you're sure it's the same hose line. **Don't lose contact with the line.** If you must move an obstacle, place the hose line under a knee or between your legs but don't lose contact with it. Quickly recheck couplings when encountered to ensure that you're moving in the right direction.

RAPID LOCATION OF A DOOR

When forced to escape, if you're not in contact with a wall then rapidly move and sweep in front of you until you contact one. Once you're on a wall, choose a direction and go! While moving along the wall make sure to sweep high enough on the wall to feel for a window, failure to do this may cause you to pass by a potential exit.

When a door is encountered decide whether to proceed through the door or bypass it. The objective is to get out of the building or get to a safe area that will buy some time. If the door

leads to the outside then the decision is obvious—get out! If it's an interior door then quickly assess the options. Does the door lead into a closet? Does the door lead to a set of stairs (up or down, what floor are you on)? What are the conditions like beyond the door? Is there a door that can be closed? By passing through a door and closing it you may buy yourself time to determine your next move while separating yourself from the fire. If you encounter a doorframe with no door then, even if you pass through it, you'll have to keep moving because you won't be able to separate yourself from the fire.

BREACHING A WALL

When you're unable to find a door or window, or conditions are so severe that you must get out of the area immediately, you may be able to breach a wall and get to the other side (or outside). Breaching a wall is a difficult task under normal conditions and it will be extremely taxing if you're being chased out of the area.

Knowing what type of construction you're dealing with will be very important. Are you in a wood frame building? Is it concrete block? Is it masonry construction? Obviously, breaching a wood-frame wall will be much different than breaching a masonry wall. Do you carry a hand tool?

While breaching a wall and getting to the other side will buy you some time you'll still have to continue searching for a way out. The breach will provide an avenue for the interior conditions to follow you. Once you're through the opening continue moving along the wall to find a door or window.

If you're on an outside wall, and unable to find a door or window, it may be possible to breach to the outside. Exterior walls will be more difficult to breach than interior walls but it may be possible to get through them.

Performing the Breach

Here's where knowing what type of construction you're dealing with is important. While it may be easy to breach through a wood-frame (or light-weight metal stud) wall, it will be very difficult to breach through a concrete wall. *Do an individual fireground size-up!* If breaching the wall is the only alternative then do it. With frame construction (drywall or plaster and lathe), choose a location to breach and use a tool to break through the wall. Penetrate all the way through the wall to ensure that there's nothing blocking the other side (failure to check the other side will result in a lot of wasted energy if an obstacle is encountered).

Once the opening is large enough to get through (you may be able to move a stud out of the way by striking it with your tool) remove any potential entanglement hazards. Lastly, check both the integrity of the floor and the environment on the other side of the opening and make your way through. *Getting through the opening may require a reduced profile maneuver with your SCBA.* Once through, follow the wall to find a way out. While the breach provides an avenue of escape (temporarily) it also provides an avenue for the interior environment to follow, so keep moving until you've reached a safe area (or the outside).

Breaching without a tool:

If you must breach through a wall and you don't have a tool, consider the *mule kick*. This technique basically involves facing away from the

wall on your hands and knees and *kicking* the wall at the breach location. An alternative to the *mule kick* is to sit facing the wall and kick into the breach location. Once an opening is made it can be expanded and the escape continued. On your side of the wall try to pull the material towards you; on the away side of the wall, push, kick, or drive the material away from you. When it's time to go through the breach, go head first. It's much easier trying to pull yourself through using your hands than it is using your feet. If you get stuck halfway through when going feet first you'll be out of luck.

EMERGENCY WINDOW EXITS

When forced to make an emergency exit through a window consider one of the following survival techniques:

- Straddle the windowsill
- Hang and drop
- Head-first ladder slide
- Rope slide

Clear the Window

Clearing the window is an integral part of any emergency window escape. When clearing the window make sure to clean out the glass completely. Clearing the win-

dow also involves removing the sash (if present) so it doesn't become an entanglement while you're halfway out. The open window will certainly intensify the interior environment, and probably draw it toward the opening, so don't chance getting stuck on your way out.

Straddle and Hang from a Windowsill

If you remain low in the window and straddle the windowsill you may be able to buy enough time, out of the intense environment, to let conditions pass. Performing a straddle over the windowsill keeps you in the lowest portion of the window and places most of your body (except one arm and leg) on the outside. The worst conditions will be found in the upper levels of the window so staying as low as

possible in the window will help reduce your exposure. If you're at a first floor window when performing this then lower yourself to the ground and get away from the window.

If you're on an upper floor and the straddle doesn't provide enough relief then consider rotating all the way out the window until only your gloved hands remain in contact with the windowsill. This will be a very difficult position to hold but it may buy you enough time for help to arrive or for interior conditions to change enough to allow you to pull yourself back into the window.

If conditions don't improve and you're still hanging from the windowsill then consider the Hang and Drop technique.

Hang and Drop

In the event that you're hanging from an upper story window (second or possibly third floor), and conditions won't allow a re-entry, then consider dropping to the ground. Any fall (or jump) from this height could result in serious injury-or worse-but there may be no other options available. By extending your

body fully while hanging from the windowsill you'll reduce the overall distance of the fall. If the second floor windowsill is fifteen feet off the ground then you'll be dropping a little less than ten feet. If you must drop to the ground then anticipate the point of impact and factor in the location of the SCBA. Try to

avoid a direct fall onto the SCBA due to the potential injuries. **The hang and drop is an extreme solution for an extreme problem!**

Head-First Ladder Slide

The head-first ladder slide may be used as an emergency escape technique when encountering a window with a ground ladder. Performing this technique puts you at great risk—but when no other option is available it may save your life. If conditions allow, exit the window and descend the ladder in a normal fashion. If interior conditions don't allow a normal window exit onto the ladder then exiting the window and sliding down the ladder in a head-first fashion will allow you to reach safety.

FIREGROUND LADDER PLACEMENT

Ladder Angle

While traditional fireground ladder placement teaches you to place ground ladders at the proper climbing angle (75°, 1/4 of the distance...), it may be wise to reexamine this placement as it relates to emergency egress from the fire building. A traditional climbing angle creates an extremely steep angle when performing the ladder slide. While the ladder slide can be accomplished at this angle, a lesser angle provides a smoother transition from the window to the ground—allowing you to perform a slower, controlled, descent. Consider a lesser angle when placing ground ladders for emergency and alternate egress from a building.

Ladder Tip

The tip of the ladder should be placed at, or slightly below, the windowsill. Placing the ladder tip at this location reduces the chance of getting hung-up on the ladder rails as you make the transition from inside to outside.

Performing the Head-First Ladder Slide

When approaching a window with a ground ladder reach over the windowsill and grab the rungs with both hands. While maintaining a low profile in the window, pull yourself up and onto the ladder and use your hands to reach, and grab, the next rung. As you continue making your way down the ladder with your hands, your feet will exit the window and provide an additional braking mechanism by contacting each rung. If you must stop and reposition, maintain a firm grip with your hands and feet and reposition so that you can continue. **When performing the head-first ladder slide it's essential to maintain control of your body throughout your entire descent.** Once you reach the bottom of the ladder quickly move to the side to allow other members, who may be coming out the window behind you, to perform the slide and safely reach the bottom.

When exiting onto the ladder, grasp the rungs and pull yourself up and onto the ladder.

When performing the head-first ladder slide you must maintain control throughout your entire descent.

Initially, hook your feet on the windowsill to transition your weight from inside to the ladder.

During the slide, use both your hands and feet to maintain control.

While the risk of injury still exists—the alternative to not performing this type of training is far more costly!

When approaching the bottom of the ladder prepare to 'roll' out of the way so others can escape.

Variations of the Ladder Slide

There are a number of variations that are taught when it comes to performing the ladder slide. The version depicted here involves a straight descent down the ladder once you've left the window.

One version that is taught involves rotating your body during the descent to place yourself in a feet-first position. The seriousness of the situation leading to the ladder slide, the potential for other firefighters performing the skill directly behind you, and the potential of the ladder slipping during the rotation are reasons why the straight descent is shown here.

Rope Slide

The rope slide is another last-ditch survival technique that may have to be used to exit an upper floor when no other alternative exists. The rope slide, like the head-first ladder slide, puts you at great risk and should be avoided if at all possible. To perform the rope slide you must have a piece of rescue rope, a way to secure it to an anchor, and an adequate anchor point.

There are a number of variations of the rope slide that are currently taught. Some versions use a personal harness and a descending device and some simply teach you to wrap the rope around your body and control the descent with your hands. Proper training is the key to using any version of this technique.

The rope slide can be broken down into three distinct steps that must be performed:

- Clear the window
- Secure an adequate anchor point and test integrity
- Rig the rope, exit the window and perform the rope slide

Clear the Window

When clearing the window make sure to clean out the glass completely. Glass and rope don't mix. If you fail to clear out the glass then it may cut the rope. Clearing the window also involves removing the sash (if present) so it doesn't become an entanglement while you're halfway out. The open window will certainly intensify the interior environment, and probably draw it toward the opening, so don't chance getting stuck on your way out. **Securing the anchor and clearing the window may be done in a reverse order, depending on where the anchor point is set (window, wall, etc.).**

Secure an Adequate Anchor

Securing an adequate anchor is a critical step in performing the rope slide. The most important feature of any anchor used is that it be capable of holding your weight during the emergency exit.

What's a good anchor?

There's no definitive list of answers. The generic (and appropriate) answer is simple—anything that will support your weight and allow you to escape. Possible anchors include hand tools (in various positions and locations), furniture, doors, wall studs or floor joists,

door hinges, or whatever else will satisfy the critical need *(being able to support your weight throughout the escape).*

One of the most secure anchor points can be created, if construction type allows it, by breaching a couple holes in a wall near a stud and securing your rope to the stud. Remember, this type of anchor will be strongest at its lowest point (closer to the floor).

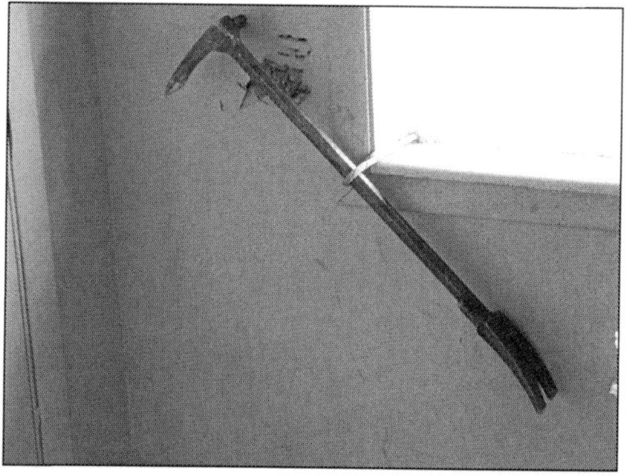

Anchor points can also be created with a hand tool, if you carry one. One method of securing an anchor point with a tool is to firmly drive the tool into the floor and attach the rope. A second method is to position the tool diagonally in one of the low corners of the window (set the tool into the corner by driving it in if possible) and attaching the rope. Training will help you identify other possible choices.

If you need to search for an anchor location away from the planned escape window, deploy the rope first. This way, if the smoke condition causes you to lose sight of the window while you're searching for an anchor, you can simply follow the rope back to the window.

Test the anchor!

Place your weight onto it before you actually use it. For example: if you're wrapping the escape rope around a halligan tool that is set into the floor then put your weight onto the rope, and anchor, as you move towards the window. If the anchor is going to fail it's likely to fail during this test (although there are no guarantees).

Training is the key!

Train with various anchor points under controlled conditions, this should be obvious. Have a training session that is designed to try various anchor points that might be found in the types of structures you respond to. During the session set up scenarios that allow you to self-escape using each of the anchor points. Your safety line will let you know if the anchor will work. The more options you try during training the more you'll have to draw from during an emergency.

Performing the Rope Slide

The version of the rope slide described here uses only a personal escape rope (refer to NFPA 1983) attached to a solid anchor, the friction which controls the descent is created by the firefighter's gloved hands and equipment. Having a tool will increase your options when it comes to securing an anchor point.

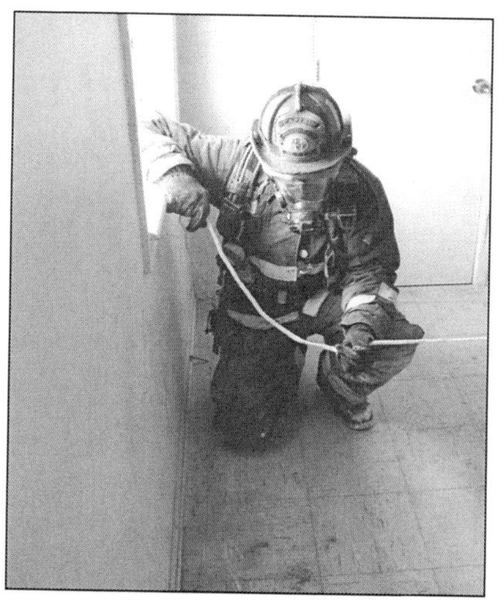

As you move from the anchor to the window keep the rope tight to remove any slack. You'll have to stay low when rigging the rope.

Rig the rope by wrapping it behind your back and under your armpits so that it comes together in the front. Grasp it tight with both hands and remove any remaining slack.

Once the anchor point is secured, and tested, keep the rope tight and extend it as you approach the window. When you're at the window and ready to exit, the rope should be wrapped under your armpits, around the back

The rope will create friction as it crosses the SCBA cylinder—adding control to the descent.

As you make the transition from inside to outside, straddle the windowsill while keeping the rope tight in your hands, and between you and the anchor. When beginning to transfer your weight onto the rope make sure not to pinch your hands between the rope and the windowsill.

Keeping the rope tight, transfer your weight from the windowsill straddle to the rope. Once your weight is on the rope system use your hand grip to control your descent to the ground.

or your body (this becomes a friction device) and brought together in front of your chest. Grasp the rope tightly in both gloved hands and take out any slack between you and the anchor. By adjusting your grip the rope will slide through your gloved hands. As you approach the windowsill you'll have to remain low—due to the interior conditions—and straddle the sill. While maintaining a firm grasp of the rope in both hands begin to lower yourself off of the windowsill, allowing your body weight to be transferred to the rope and anchor. Once you've transferred your weight and your body is outside of the window, begin to slowly lower yourself to safety by adjusting your hands to allow the rope to slide. Safety can be the ground, a porch roof, or a floor below (re-enter through a window and exit the building by the interior stairs).

Note: As the rope travels behind your back it may slide across the SCBA cylinder or it may slip down and ride under the cylinder and harness assemble, against your back. Either position is acceptable because friction will be generated in both places, helping to maintain control during your descent.

Performing the rope slide with a Rescue Harness

When using a rescue harness with a descent device you'll have to rig the rope differently. The rope must be properly rigged into the descent device before exiting the window. Once your body weight is transferred to the rope and anchor you control your descent by operating the descent device. **Rescue rope, harness and descent device selection:** For information on the selection of personal escape rope, a rescue harness, and a descent device you should consult the National Fire Protection Association Standard 1983, *Fire Service Life Safety Rope and System Components.*

PREVENTION IS THE KEY TO FIREFIGHTER SURVIVAL

IF FACED WITH A FIREGROUND EMERGENCY:

Stay Low!

Stay Calm!

Stay Oriented!

NEVER GIVE UP!

7

CHAPTER SEVEN

Firefighter Survival Training Sessions

There are a number of firefighter survival skills that can be done during company training sessions. Training sessions can be as simple as donning an SCBA or they may involve breathing an entire SCBA cylinder down and performing an emergency out-of-air escape. What's important is that you continually perform survival training—*give yourself a fighting chance when faced with a true fireground emergency.*

The following training sessions should focus on both individual skills and complete MAYDAY situations. Once you're comfortable performing the skill then practice it in the *fireground context,* the conditions you're likely to face during a true emergency.

Fireground context includes:

- Full personal protective clothing
  - That means all of your turnout gear.
  - Firefighting gloves, not leather work gloves.
  - A hood on and in place
- SCBA with PASS device armed
  - Breathing off of the SCBA, not just wearing it.
  - If the PASS device chirps, move.
- Blacked-out SCBA face piece
- Realistic fireground noise
  - Water flowing, glass breaking, smoke detectors beeping, radio transmissions, engines running, fans operating, and any other fireground sounds you're likely to encounter.
- Realistic fireground activities

  While everyone can't be performing the survival skills at the same time, make them review fireground basics—hose stretches, search and rescue, ground ladder placement, horizontal and vertical ventilation, pumping, operating the aerial, and anything else they may need to do on the fireground.

TRAINING IS THE KEY TO SURVIVAL!

PREPARING FOR SURVIVAL DRILLS

Personal Equipment Review

Have each member of the company don their protective gear and give a *show-and-tell* of the equipment that they carry. Make sure they explain why they carry what they do.

Apparatus Review

Divide the apparatus up into as many sections/areas as there are members in the company. Have each member go over their assigned area of the apparatus. Have them describe the equipment—both its uses and maintenance procedures.

District Review

Have each member of the company give a brief *individual fireground size-up* of their own house. Once that's done, have each member give an individual size-up of a potential problem occupancy in the response district.

MANAGING YOUR MAYDAY DRILLS

MAYDAY Transmission

Have each member of the company perform a search and rescue of some type of occupancy (whatever you have available). While performing the search in blacked-out mask have the firefighter transmit a MAY-

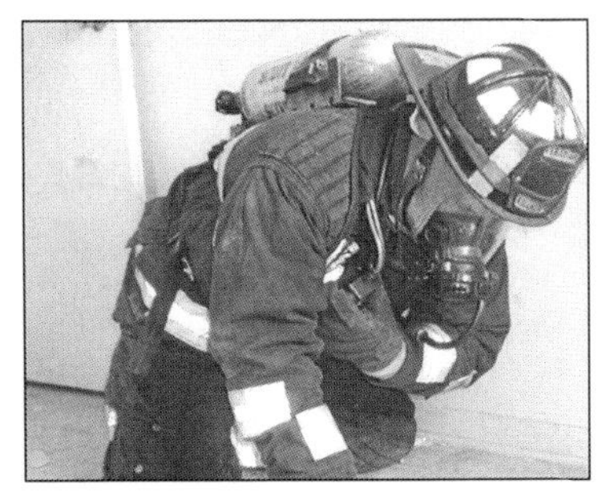

DAY over the radio. Review LUNAR and make sure they provide as much information as possible about their problem and location in the building.

PASS Activation

While wearing full turnout gear, and a blacked-out mask, have each member place their PASS device into manual alarm mode. Gloved hands!

SCBA EMERGENCY DRILLS

Create a simple SCBA emergency training prop. The prop can be multi-purpose and allow both reduced profile and entanglement training sessions. A couple pieces of plywood, a few 2 x 4's, a saw, hammer and nails is all it a takes. Once you've made it you can use it over-and-over again to keep your skills sharp.

SCBA Familiarization

Take an SCBA unit into a closed room, partially disassemble the unit by unthreading the high pressure hose from the cylinder. Twist various straps around one another, loosen the cylinder from the harness, and activate the bypass and regulator. Leave the room in total darkness. Have each member enter the room, with full gear on, find the SCBA and put it on and into service. **Requirement:** No air leaks during the operation, no twisted straps when complete, the unit must be assembled and donned correctly when they exit the room.

Reduced Profile Maneuvers

Have each member perform a partial and complete removal reduced profile maneuver. Make sure to use a blacked-out mask.

Entanglement Emergencies

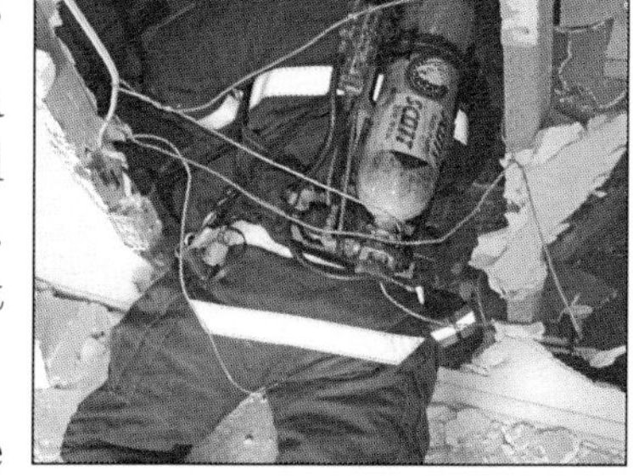

Have each member perform a partial and complete removal to remove an entanglement. Make sure to use a blacked-out mask.

Once they're proficient at the partial and complete removal techniques, advance their training by requiring the use of wire cutters to remove the entanlgment.

Equipment Failures

Gather all members together and let them watch, and time, as you open the cylinder valve of an SCBA cylinder and let it completely empty. ***How long does it take?***

Next, while members are performing interior, blacked-out,

operations, walk behind them with an SCBA cylinder and crack open the cylinder valve. Leave the cylinder open until they reach for, and begin to operate, their SCBA cylinder valve–*controlling their air supply.*

High Pressure Leak – Controlled at Cylinder

Repeating the above interior operation with blacked-mask, approach each company member and instruct them they have a high pressure leak (with no compromise of air to the mask). Have them escape the building while controlling their breathing with the cylinder valve.

Out-of-Air Emergencies

During a blacked-out interior operation have each member simulate an out-of-air emergency and exit the building—keeping low and filtering the air with a gloved-hand.

DISORIENTATION EMERGENCY DRILLS

Find and Follow a Wall

Place individuals inside a large room with at least one door. Disorient them in the room (wheel them in on a chair, lead them through the room in a blacked-out mask and spin them around before they begin...) and have them call for help and find their way out. Make sure to have them alarm their PASS device.

Pile Of Hose

Advance two or three hose lines into a structure. While advancing the lines make sure to go around and over/under objects and cross the lines multiple times. Make sure to put a few loops in the lines as well. Again, with blacked-out masks, lead firefighters into the building and disoriented them somewhere near a hose line. Tell them to call for help (MAYDAY), find a line and make their way out.

Advanced Get Out Drills

Lead firefighters, as companies, into and through different occupancy types while wearing black-out masks. Make sure to have a few hose lines advanced into the structure to provide some potential assistance during their escape. Somewhere inside the structure separate the company, disorient them further, have them alarm their PASS devices, and find a way out. ***Do they work together? Does anybody take charge?***

EMERGENCY ESCAPE TECHNIQUES

Breach a Wall

Construct a simple wall prop out of 2 x 4's, drywall, insulation and wires. Have each member breach through the wall, first with a tool and then without one. Perform a reduced profile maneuver to get to the other side.

Straddle and Exit a First-Floor Window

Disorient each member in a first-floor room with a window and a door. Lock the door and have them escape the room by clearing the window, straddling the windowsill, and lowering themselves to safety.

IF YOU HAVE NOT FORMALLY TAKEN A COURSE THAT TEACHES THE HEAD-FIRST LADDER SLIDE OR THE ROPE SLIDE, SEEK ONE OUT AND GET THE TRAINING—YOUR LIFE MAY DEPEND ON IT!

Head-First Ladder Slide and Rope Slide

With proper training, and all safety measures in place, perform emergency window escape training in both the head-first ladder slide and the rope slide.

Remember, part of being able to perform these skills during a true emergency invovles having the equipment when you need it. Work towards being proactive on the fireground by laddering the building for emergency escape.

Ground Ladders

Place ground ladders to different window locations. Make sure to adjust the angle for emergency escape.

Rope Slide Anchors

Practice setting and testing a variety of anchors that may be used when performing the rope slide to escape an untenable environment.

NOTES

NOTES

NOTES

NOTES

NOTES

NOTES

NOTES

NOTES